I0467874

NISTIR 7956

Cryptographic Key Management Issues & Challenges in Cloud Services

Ramaswamy Chandramouli
Michaela Iorga
Computer Security Division
Information Technology Laboratory

Santosh Chokhani
Cygnacom Solutions

September 2013

U.S. Department of Commerce
Penny Pritzker, Secretary

National Institute of Standards and Technology
Patrick D. Gallagher, Under Secretary of Commerce for Standards and Technology and Director

Certain commercial entities, equipment, or materials may be identified in this document in order to describe an experimental procedure or concept adequately. Such identification is not intended to imply recommendation or endorsement by NIST, nor is it intended to imply that the entities, materials, or equipment are necessarily the best available for the purpose.

There may be references in this publication to other publications currently under development by NIST in accordance with its assigned statutory responsibilities. The information in this publication, including concepts and methodologies, may be used by Federal agencies even before the completion of such companion publications. Thus, until each publication is completed, current requirements, guidelines, and procedures, where they exist, remain operative. For planning and transition purposes, Federal agencies may wish to closely follow the development of these new publications by NIST.

Organizations are encouraged to review all draft publications during public comment periods and provide feedback to NIST. All NIST Computer Security Division publications, other than the ones noted above, are available at http://csrc nist.gov/publications.

Reports on Computer Systems Technology

The Information Technology Laboratory (ITL) at the National Institute of Standards and Technology (NIST) promotes the U.S. economy and public welfare by providing technical leadership for the Nation's measurement and standards infrastructure. ITL develops tests, test methods, reference data, proof of concept implementations, and technical analyses to advance the development and productive use of information technology. ITL's responsibilities include the development of management, administrative, technical, and physical standards and guidelines for the cost-effective security and privacy of other than national security-related information in Federal information systems.

Abstract

To interact with various services in the cloud and to store the data generated/processed by those services, several security capabilities are required. Based on a core set of features in the three common cloud services - Infrastructure as a Service (IaaS), Platform as a Service (PaaS) and Software as a Service (SaaS), we identify a set of security capabilities needed to exercise those features and the cryptographic operations they entail. An analysis of the common state of practice of the cryptographic operations that provide those security capabilities reveals that the management of cryptographic keys takes on an additional complexity in cloud environments compared to enterprise IT environments due to: (a) difference in ownership (between cloud Consumers and cloud Providers) and (b) control of infrastructures on which both the Key Management System (KMS) and protected resources are located. This document identifies the cryptographic key management challenges in the context of architectural solutions that are commonly deployed to perform those cryptographic operations.

Keywords

Authentication; Cloud Services; Data Protection; Encryption; Key Management System (KMS); Secure Shell (SSH); Transport Layer Security (TLS)

Table of Contents

Executive Summary...1

1. Cryptographic Key Management Overview 2

1.1   Key Types 2

1.2   Key States..4

1.3   Key Management Functions 5

1.4   Key Management - Generic Security Requirements..7

2. Cloud Computing Environment – Evolution & State of Practice..8

2.1   Three Generations of Internet 8

2.2   Cloud Computing Definition (by NIST) 8

2.3   Cloud Computing Reference Architecture (from NIST)..10

3. Cryptographic Key Management Challenges in the Cloud 15

3.1   Challenges in Cryptographic Operations & Key Management for IaaS 16

3.2   Challenges in Cryptographic Operations and Key Management for PaaS.............................22

3.3   Challenges in Cryptographic Operations & Key Management for SaaS 22

Appendix A – Security Analysis of Cryptographic Techniques for Authenticating VM Templates in the Cloud 25

A.1. VM Template Authentication using Digital Signature 25

A.2. VM Template Authentication using Cryptographic Hash Function 26

A.3. VM Template Authentication using Message Authentication Code (MAC) 28

A.4. VM Template Authentication Based on Cloud Provider Discretionary Access Control...........30

A.5. Conclusion..30

Appendix B - Bibliography 31

Executive Summary

Encryption and access control are the two primary means for ensuring data confidentiality in any IT environment. In situations where encryption is used as a data confidentiality assurance measure, the management of cryptographic keys is a critical and challenging security management function, especially in large enterprise data centers, due to sheer volume and data distribution (in different physical and logical storage media), and the consequent number of cryptographic keys. This function becomes more complex in the case of a cloud environment, where the physical and logical control of resources (both computing and networking) is split between cloud actors (e.g., Consumers, Providers, and Brokers) (see Section 2.2 below and NIST SP 500-292 for more details).

The objectives of this document are to identify:

(a) The cryptographic key management issues that arise due to the distributed nature of IT resources, as well the distributed nature of their control, the latter split among multiple cloud actors. Furthermore, the pattern of distribution varies with the type of service offering - Infrastructure as a Service (IaaS), Platform as a Service (PaaS) and Software as a Service (SaaS).

(b) the special challenges involved in deploying cryptographic key management functions that meet the security requirements of the cloud Consumers, depending upon the nature of the service and the type of data generated/processed/stored by the service features.

In this document, we address the following topics:

1. Section 1 provides an overview of cryptographic key management;
2. Section 2 provides a summary of the cloud computing concepts, including a reference architecture (cloud actors, cloud service types and deployment models) as identified in NIST standards; and
3. Section 3 builds on the previous sections to identify a core set of features for the three main cloud service types – IaaS, PaaS and SaaS: the security capabilities (SC) required to exercise those features, architectural solutions available to meet the security capabilities and the consequent key management challenges.

In order to ensure that cryptographic mechanisms provide the desired security, the following criteria should be met with regards to their three main components – Algorithms (and associated modes of operation), Protocols and Implementation:

1. The cryptographic algorithms and associated modes of operation deployed should have been scrutinized, evaluated, and approved using a review process that is open and includes a wide range of experts in the field. Examples of such approved algorithms and

modes are found in National Institute of Standards and Technology's Federal Information Processing Standards (FIPS) and Special Publications (SPs), and in the Internet Engineering Task Force (IETF) Request for Comment (RFC) documents. The specific NIST documents pertaining to cryptographic algorithms and associated modes of operation are: FIPS 186-3 for Digital Signatures, FIPS 180-4 for Secure Hash, SP 800-38A for modes of operation and SP 800-56A & SP 800-56B for key establishment.

2. The cryptographic protocols used should have been scrutinized, evaluated, and approved using a review process that is open and includes a wide range of experts in the field. IETF protocol specifications for Secure Shell (SSH) and Transport Layer Security (TLS) are examples that meet these criteria.

3. The implementation of a cryptographic algorithm or protocol should undergo a widely recognized and reputable independent testing for verification of conformance to underlying specifications. NIST's Cryptographic Algorithm Validation Program (CAVP) and Cryptographic Module Validation Program (CMVP) are examples of such independent testing programs.

1. Cryptographic Key Management Overview

In this section, we review the two broad categories of cryptographic keys, list the most commonly used key types, identify the key states and chart the resulting transition diagram. We then describe the most important key management functions (also referred to as key lifecycle operations) and list the generic security requirements associated with these functions.

1.1 Key Types

Cryptographic keys fall into two broad categories:

1. **Secret key:** A key that is generally used to 1) perform encryption/decryption using symmetric cryptographic algorithms; and/or 2) to provide data integrity using message authentication codes (i.e., Hash based Message Authentication Code or HMAC) or an encryption mode of operation that also provides data integrity. A secret key is also called a symmetric key, since the same key is required for encryption and decryption or for integrity value generation and integrity verification.

2. **Public/Private Key Pair:** A pair of mathematically related keys used in asymmetric cryptography for authentication, digital signature, or key establishment. As the name indicates, the private key is used by the owner of the key pair, is kept secret, and should be protected at all times, while the public key can be published and used be the relying party to complete the protocol or invert the operations performed with the private key.

From these broad categories one can determine the most commonly used key types in a cloud computing environment. This is not to say that a cloud implementation may not have additional types of keys.

1. **Public/Private Authentication Key Pair:** This key pair is used by one party (peer, client or server) to authenticate to the other party. Its typical use entails combining a random challenge with the signer-generated random number and signing the result for the benefit of the challenger who wishes to authenticate the private-key holder. Examples of usage include client-authenticated Transport Layer Security (TLS), Virtual Private Network (VPN) authentication, and smart card-based logon. An authentication key pair is generally used in a network environment and is generally used for long-term use (e.g., up to 3 years)

2. **Public/Private Signature Key Pair:** The private key of the key pair is used by one party to digitally sign a message/data, while the corresponding public key is used to verify the signature. Examples of the usage of a signature key pair are signed Secure/Multipart Internet Mail Extensions (S/MIME) messages, signed electronic documents, and signed code. In some implementations, a key pair may be used for both authentication and signature functions. A signature key pair is generally used in a network environment and is generally used for long-term use (e.g., up to 3 years). It may also be used to generate and verify signatures on stored data.

3. **Public/Private Key Establishment Pair:** This key pair is used to securely establish a key between parties. Examples of the use of a key pair for key establishment are encrypting the symmetric key for S/MIME payload encryption/decryption and encrypting the random secret to be sent from a TLS client to a server. It is recommended that key establishment key pairs be distinct from authentication and signature key pairs. However, it is recognized that some devices such as web servers use the same key pair for key establishment and authentication. A key establishment key pair is traditionally used in a network environment, but some usage for stored data is also seen and can be envisioned. A key establishment key pair is generally used for a pre-defined period for encryption (e.g., up to 3 years), but is used for decryption for as long as the confidentiality of the data needs to be protected.

4. **Symmetric Encryption/Decryption Key:** A symmetric key is used to encrypt and decrypt data or messages. For data-in-transit, a symmetric encryption/decryption key may have a short life, typically for each message (e.g., S/MIME message) or for each session (for example a TLS session). For stored data, the symmetric life of the encryption/decryption key tends to be as long as the confidentiality of the data needs to be protected.

5. **Symmetric Message Authentication Code (MAC) Key:** A symmetric key is used to provide assurance for the integrity of data. There are three techniques used to provide this assurance: 1) use a symmetric encryption algorithm and a MAC mode of operation (e.g., CMAC using AES); 2) use a symmetric encryption algorithm and an authenticated encryption mode of operation (e.g., GCM or CCM using AES); and 3) use a hash-based MAC (HMAC). For data-in-transit, a symmetric MAC key has a short life, typically for a single message or for a single session (for example a TLS session). For stored data, the life of a symmetric MAC key tends to be for as long as the data needs to be protected. Note that when authenticated encryption mode is used, the same key is used for both the

MAC and encryption/decryption, since both objectives are achieved by invoking a single mode of operation.

6. **Symmetric Key Wrapping Key:** A symmetric key is used to encrypt a symmetric key or an asymmetric private key. A Key Wrapping Key is also called a Key Encrypting Key.

1.2 Key States

A symmetric key or public/private key pair can undergo the following states. This is not to say that a key management implementation may not have additional states. Alternatively, a key management implementation may have a subset of these states.

- **Generation:** A symmetric key or public/private key pair is generated when required.

- **Activation:** A symmetric key or private key is activated when it is required to be used. A public key is activated when it is made available or on the date indicated in its associated metadata (e.g., notBefore date in an X.509 public key certificate).

- **Deactivation:** A symmetric key or private key is deactivated when it is no longer required for applying cryptographic protection to data. Deactivation of these keys may be followed by destruction or archival. A public key is not deactivated. It may expire (e.g., at the notAfter date in an X.509 public key certificate), or may be suspended (e.g., via certificate revocation list (CRL) [refer RFC 4949] in X.509 standard) or revoked (e.g., via CRL in X.509 standard).

- **Suspension:** A key may be suspended from use for a variety of reasons, such as an unknown status of the key or due to the key owner being temporarily away. In the case of the public key, suspension of the companion private key is communicated to the relying parties. This may be communicated as an "On hold" revocation reason code in a CRL and in an Online Certificate Status Protocol (OCSP) response

- **Expiration:** A key may expire due to the end of its crypto period [refer RFC 4949]. In the case of a public key, an expiration date is indicated in the associated metadata (e.g., notAfter date in X.509 certificates).

- **Destruction:** A key is destroyed when it is no longer needed.

- **Archival:** A key may be archived when it is no longer required for normal use, but may be needed after the key's cryptoperiod. An example for secret or private keys is the possible decryption of archived data. An example for public keys is the verification of archived signed documents.

- **Revocation:** A revocation is explicitly stated with respect to public keys; however, the revocation also applies to the corresponding private key. Revocation information is securely communicated to the relying parties, for example, as CRLs or OCSP responses, in the case of X.509 public key certificates. Secret keys are also "revoked", often by including them on lists, such as a compromised key list.

The following is the state diagram for the key states.

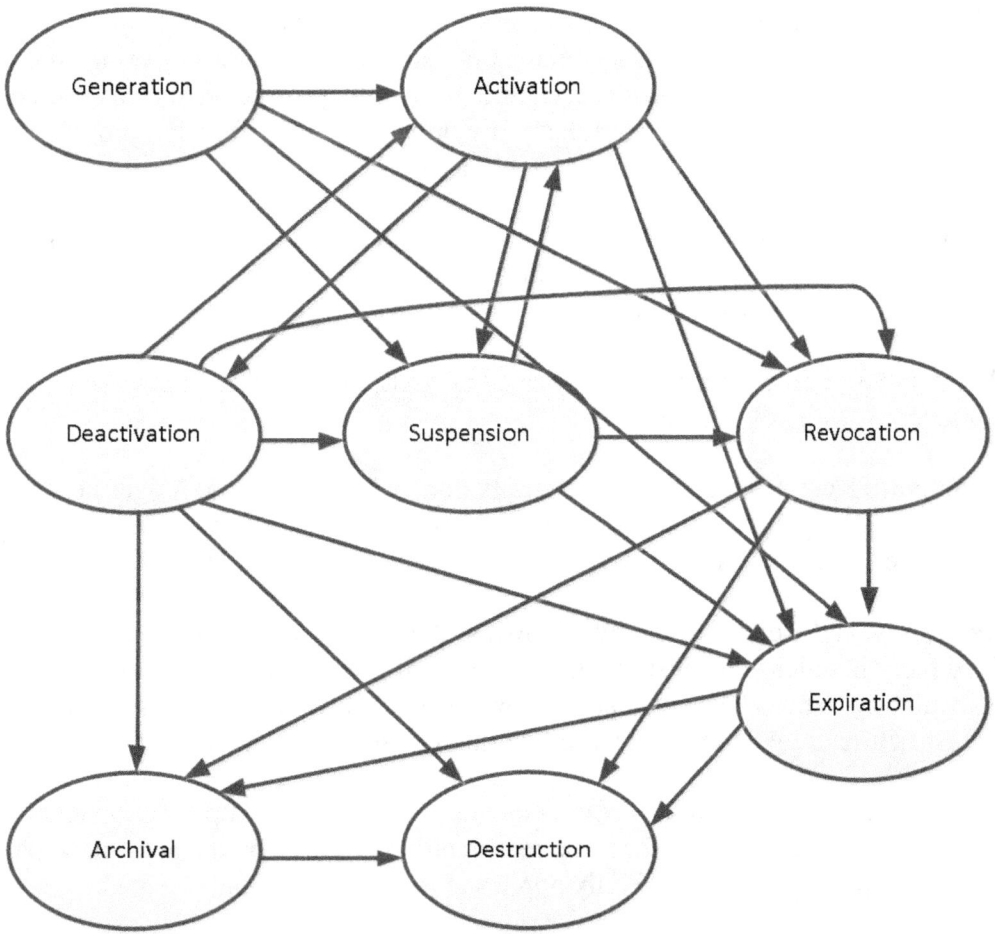

Figure 1: State Diagram for the Key States

1.3 Key Management Functions

The following are the important key management functions:

- **Generate Key:** The generation of good-quality keys is critical to security. Keys for a cryptographic algorithm should be generated in cryptographic modules that have been approved for the generation of keys for that algorithm.

- **Generate Domain Parameters:** Discrete Logarithm-based algorithms require the generation of domain parameters prior to the generation of the keys; the keys are generated using those domain parameters. The domain parameters for an algorithm shall be generated in approved cryptographic modules that have been approved for their generation. Since domain parameters can be common to a broad community of users, key generation need not entail domain parameter generation. For example, defining Suite B P-256 curve defines all the domain parameters for the attendant ECDSA and ECDH algorithms.

- **Bind Key and Metadata:** A key may have associated data, such as the time period of use, usage constraints (such as authentication, encryption, and/or key establishment), domain parameters, and security services for which they are used, such as source authentication, integrity, and confidentiality protection. This function provides assurance that the key is associated with the correct metadata.

- **Bind a Key to an Individual:** The identifier of the individual or other entity that owns a key is considered as part of the key's metadata, but this association is sufficiently critical to be listed as a distinct function.

- **Activate Key:** This function transitions a key to the active state. It is often done in conjunction with key generation.

- **Deactivate Key:** This function is generally done when a key is no longer needed for applying cryptographic protection. For example, when a key has expired, or is replaced by another key.

- **Backup Key:** A key is backed by the owner, the key management infrastructure, or a third party in order to reconstitute the key when it is accidentally destroyed or otherwise unavailable. When a private or secret key is backed up by the key management infrastructure or by a third party, the function is also referred to as "key escrow".

- **Recover Key:** This function is complementary to the key backup function and is invoked when the key is unavailable for some reason and is required by the authorized parties. Key backup and recovery generally applies to the symmetric and private keys.

- **Modify Metadata:** This function is invoked when metadata bound to a key needs to change. The renewal of a public key certificate is an example of this function where the validity period for the public key is changed.

- **Rekey:** This function is used to replace the existing key with a new key. Generally, the existing key (the key being replaced) plays a role in authentication and authorization for replacement.

- **Suspend a Key:** This function is used to temporarily cease the use of a key. It is akin to reversible revocation. This function may need to be invoked if the status of a key is undetermined or if the key owner wishes to temporarily suspend its use (e.g., for extended leave). For secret keys, this can also be accomplished via key deactivation. For public keys and the companion private key, this is generally done using suspension notification of the public key.

- **Restore a Key:** This function is used to restore a suspended key once its secure status is ascertained. For secret keys, this can also be accomplished via key activation. For public keys and the companion private keys, this is generally done using a revocation notification where the revoked public key entry is deleted implying the key is valid.

- **Revoke a Key:** This function is used to inform the relying parties to stop using a public key. There may be a variety of reasons for this, including the compromise of companion private key, and the owner having stopped using the companion private key.

- **Archive a Key:** This function is used to store a key in long-term storage after it has been deactivated, expired, and/or compromised.

- **Destroy a Key:** This function is used to zeroize a key when it should no longer be used.

- **Manage TA Store:** This function is used by the relying party to determine what trust anchors to trust for what purpose. A trust anchor is a public key and its associated metadata that the relying party explicitly trusts and uses to establish trust in other public keys via transitive trust, such as a public-key certification path that is a series of public key certificates where the digital signature in one certificate can be used to verify the digital signature on the next certificate.

1.4 Key Management - Generic Security Requirements

The following are general key management security requirements:

1. Parties performing key management functions are properly authenticated and their authorizations to perform the key management functions for a given key are properly verified.

2. All key management commands and associated data are protected from spoofing, i.e., source authentication is performed prior to executing a command.

3. All key management commands and associated data are protected from undetected, unauthorized modifications, i.e., integrity protection is provided.

4. Secret and private keys are protected from unauthorized disclosure.

5. All keys and metadata are protected from spoofing, i.e., source authentication is performed prior to accessing keys and metadata.

6. All keys and metadata are protected from undetected, unauthorized modifications, i.e., integrity protection is provided.

7. When cryptography is used as a protection mechanism for any of the above, the security strength of the cryptographic mechanism used is at least as strong as the security strength required for the keys being managed.

There are significant challenges to implementing these key management security requirements in cloud computing over unsecure public networks. In the next sections, we review the cloud computing reference architecture and identify, for the three main cloud service types – IaaS, PaaS and SaaS - a core set of features, the security capabilities (SC) required to exercise these features, architectural solutions available to meet the security capabilities, and the consequent key management challenges.

2. Cloud Computing Environment – Evolution & State of Practice

2.1 Three Generations of Internet

The evolution of the internet can be divided into three generations: in the 70s the first generation was marked by expensive mainframe computers accessed from terminals; the second generation was born in the late 80s and early 90s and was identified by the explosion of personal computers with Graphical User Interfaces (GUIs); the first decade of the 21st century brought the third generation, defined by mobile computing, the "internet of things" and cloud computing.

In 1997, Professor Ramnath Chellappa of Emory University, defined cloud computing for the first time while a faculty member at the University of South California, as an important new *"computing paradigm where the boundaries of computing will be determined by economic rationale rather than technical limits alone."* Even though the international IT literature and media have come forward since then with a large number of definitions, models and architectures for cloud computing, autonomic and utility computing were the foundations of what the community commonly referred to as "cloud computing". In the early 2000s, companies started rapidly adopting this concept upon the realization that cloud computing could benefit both the Providers as well as the Consumers of services. Businesses started delivering computing functionality via the Internet, enterprise-level applications, web-based retail services, document-sharing capabilities and fully-hosted IT platforms, to mention only a few cloud computing use cases of the 2000s. The latest widespread adoption of virtualization and of service-oriented architecture (SOA) promulgated cloud computing as a fundamental and increasingly important part of any delivery and critical-mission strategy, enabling existing and new products and services to be offered and consumed more efficiently, conveniently and securely. Not surprisingly, cloud computing became one of the hottest trends in the IT armory, with a unique and complementary set of properties, such as elasticity, resiliency, rapid provisioning, and multi-tenancy.

2.2 Cloud Computing Definition (by NIST)

Cloud computing is a model for enabling convenient, on-demand network access to a shared pool of configurable resources (e.g., networks, servers, storage, applications, and services) that can be rapidly provisioned and released with minimal management efforts or service provider interaction. Enterprises can use these resources to develop, host, and run services and applications on demand in a flexible manner in any devices, anytime, and anywhere. According to the U.S. National Institute of Standards and Technology's (NIST) definition published in the NIST Special Publication SP 800-145, "cloud computing is a model for enabling ubiquitous, convenient, on-demand network access to a shared pool of configurable computing resources (e.g., networks, servers, storage, applications and services) that can be rapidly provisioned and released with minimal management effort or service provider interaction." This definition is widely accepted as a valuable contribution toward providing a clear understanding of cloud computing technologies and cloud services and it has been submitted as the U.S. contribution for an International standardization[1].

The NIST definition also provides a unifying view of five essential characteristics that all cloud services exhibit: *on-demand self-service, broad network access, resource pooling, rapid elasticity,* and *measured service*. Furthermore, NIST identifies a simple and unambiguous taxonomy of three "service models" available to cloud Consumers (Infrastructure-as-a-Service (IaaS), Platform-as-a Service (PaaS), Software-as-a-Service (SaaS)) and four "cloud deployment modes" (Public, Private, Community, and Hybrid) that together categorize ways to deliver cloud services. Since the cloud service model is an important architectural factor when discussing key managements aspects in a cloud environment, we are reproducing below the definitions for the service models provided by NIST in SP 800-145, "The NIST definition of Cloud Computing":

1. *Infrastructure as a Service (IaaS) - The capability provided to the Consumer is to provision processing, storage, networks, and other fundamental computing resources where the Consumer is able to deploy and run arbitrary software, which can include operating systems and applications. The Consumer does not manage or control the underlying cloud infrastructure, but has control over operating systems, storage, deployed applications, and possibly limited control of select networking components (e.g., host firewalls).*

2. *Platform as a Service (PaaS) - The capability provided to the Consumer is to deploy Consumer-created or acquired applications onto the cloud infrastructure that are created using programming languages and tools supported by the Provider. The Consumer does not manage or control the underlying cloud infrastructure, including network, servers, operating systems, or storage, but has control over the deployed applications and possibly the application-hosting environment configurations.*

3. *Software as a Service (SaaS) - The capability provided to the Consumer is to use the Provider's applications running on a cloud infrastructure. The applications are accessible from various client devices through a thin client interface, such as a web browser (e.g., web-based email). The Consumer does not manage or control the underlying cloud infrastructure, including network, servers, operating systems, storage, or even individual application capabilities, with the possible exception of limited user-specific application-configuration settings.*

IaaS allows cloud Consumers to run any operating systems and applications of their choice on the hardware and resource abstraction layer (hypervisors) furnished by the cloud Provider. A

---

[1] http://www.nist.gov/itl/csd/cloud-102511.cfm

Consumer's operating systems and applications can be migrated to the cloud Provider's hardware, potentially replacing a company's data center infrastructure.

PaaS allows Consumers to create their own cloud applications. Basically, the cloud Provider renders a virtualized environment and a set of tools to allow the creation of new web applications. The Cloud Provider also furnishes the hardware, operating systems and commonly used system software and applications, such as DBMS, Web Server, etc.

SaaS allows cloud Consumers to run online applications. Off-the-shelf applications are accessed over the Internet. The cloud Provider owns the applications, and the Consumers are authorized to use them in accordance with a Service Agreement signed between parties.

Cloud computing provides a convenient, on-demand way to access a shared pool of configurable resources (e.g., networks, servers, storage, applications, and services), which enables users to develop, host and run services and applications on demand in a flexible manner in any devices, anytime, and anywhere. Cloud services are those services that are expressed, delivered and consumed over a public network, a private network or in some combination (community or hybrid). These services are usually delivered in one of the following service categories identified by NIST: IaaS, PaaS and SaaS. Cloud Provider and Broker may also identify other Categories of services (such as Network-as-a-Service, Storage-as-a-Service, carrier-as-a-Service) that are practical components already embedded in the service models identified by NIST, and are not stand-alone service models that identify particular cloud architectures. Some cloud Providers might provide abstracted hardware and software resources that may be offered as a service. This allows customers and partners to develop and deploy new applications that can be configured and used remotely. Leveraging cloud services that provide opportunities to provision resources elastically enables enterprises to launch or change their business quickly and easily as needed.

2.3 Cloud Computing Reference Architecture (from NIST)

In the Special Publication SP 500-292, NIST has published the NIST Cloud Computing Reference Architecture[2] (RA). This architecture is a logical extension of the NIST cloud computing definition. It is a generic high-level conceptual model that is an effective tool for discussing the requirements, structures, and operations of cloud computing. The model is not tied to any specific vendor products, services or reference implementation, nor does it provide prescriptive solutions. The RA defines a set of cloud Actors and their activities and functions that can be used in the process of orchestrating a cloud Ecosystem. The Cloud Computing RA relates to a companion cloud computing taxonomy and contains a set of views and descriptions that are the basis for discussing the characteristics, uses and standards for cloud computing. The Actor-based model is intended to serve the expectations of the stakeholders by allowing them to

---

[2] http://collaborate.nist.gov/twiki-cloud-computing/pub/CloudComputing/ReferenceArchitectureTaxonomy/NIST_SP_500-292_-_090611.pdf

understand the overall view of roles and responsibilities in order to assess and manage the risk by implementing adequate security controls.

The NIST Reference Architecture is intended to facilitate the understanding of the operational intricacies in cloud computing. It does not represent the system architecture of a specific cloud computing system; instead, it is a tool for describing, discussing, and developing a system-specific architecture using a common framework of reference.

As shown in **Figure 2** this architecture outlines the five major cloud Actors; Consumer, Provider, Broker, Carrier and Auditor.

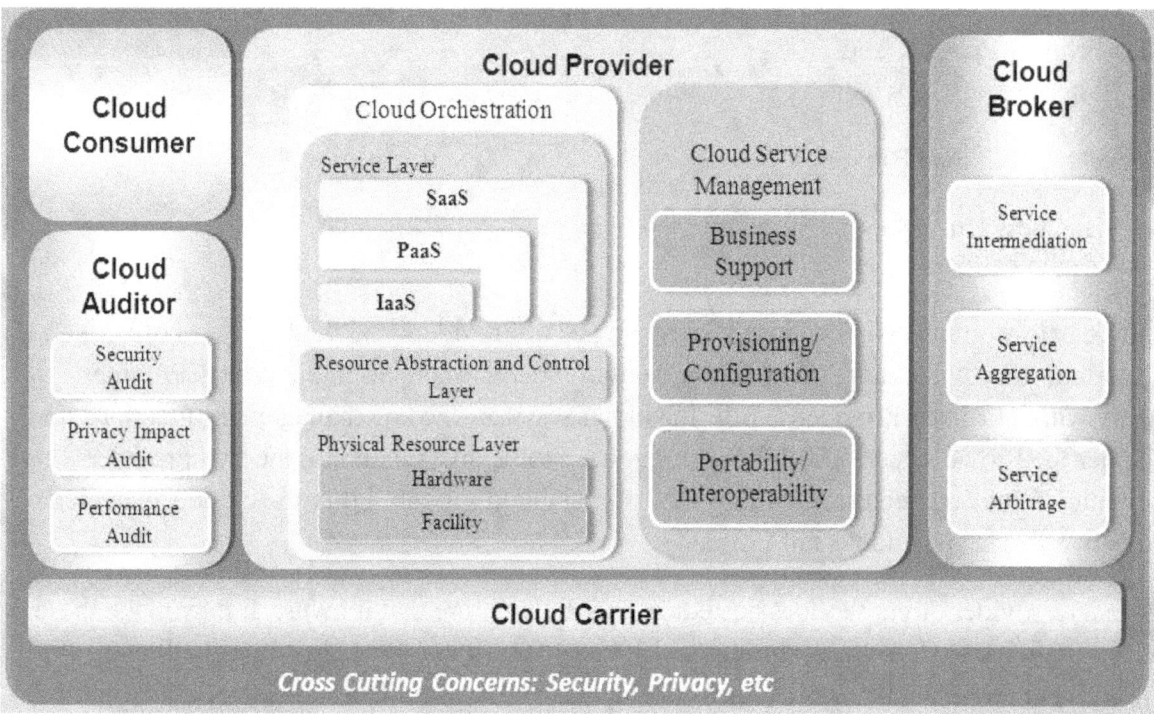

Figure 2 NIST Cloud Computing Security Reference Architecture Approach (courtesy of NIST, SP 500-292)

Each cloud Actor defined by the NIST RA is an entity (a person or an organization) that participates in a transaction or process and/or performs tasks in cloud computing. The definitions of the cloud Actors introduced by NIST in SP 500-292, NIST cloud Computing Reference Architecture, are reproduced below in Table 1.

Actor	Definition
Cloud Consumer	A person or organization that maintains a business relationship with, and uses service from, *Cloud Providers*.
Cloud Provider	A person, organization, or entity responsible for making a service available to interested parties.
Cloud Auditor	A party that can conduct independent assessment of cloud services, information

11

	system operations, performance and security of the cloud implementation.
Cloud Broker	An entity that manages the use, performance and delivery of cloud services, and negotiates relationships between *Cloud Providers* and *Cloud Consumers*.
Cloud Carrier	An intermediary that provides connectivity and transport of cloud services from *Cloud Providers* to *Cloud Consumers*.

Table 1 Cloud Actor Definitions (courtesy of NIST, SP 500-292)

In our latest work (draft documents and white papers), NIST identifies two types of cloud Providers:

1. Primary Provider and
2. Intermediary Provider,

and two types of cloud Brokers:

1. Business Broker and
2. Technical Broker.

Figure 3, below, graphically depicts these two types of Providers and the two types of Brokers. It is important to note that, in particular, cloud environments where an Intermediary Provider partners with a Primary Provider in offering cloud services, the key management functions that fall under the Provider's responsibilities might need to be divided among the two Providers, depending on the architectural details of the offered cloud service. From the cloud Consumer's perspective this segregation is not visible.

A Primary Provider offers services hosted on an infrastructure that it owns. It may make these services available to Consumers through a third party (such as a Broker or Intermediary Provider), but the defining characteristic of a Primary Provider is that it does not obtain the sources of its service offerings from other Providers.

An Intermediary Provider has the capability to interact with other cloud Providers without offering visibility or transparency into the Primary Provider(s). An Intermediary Provider uses services offered by a Primary Provider as invisible components of its own service, which it presents to the customer as an integrated offering. From a security perspective, all security services and components required of a Primary Provider are also required of an Intermediary Provider.

A Business Broker only provides business and relationship services, and does not have any contact with the cloud Consumer's data, operations, or artifacts (e.g., images, volumes, firewalls) in the cloud. The Business Broker therefore, has no responsibilities in implementing any key management functions, regardless of the cloud architecture. Conversely, a Technical Broker *does* interact with a Consumer's assets; the Technical Broker aggregates services from multiple cloud Providers and adds a layer of technical functionality by addressing single-point-of-entry and interoperability issues.

There are two key defining features of a cloud Technical Broker that are distinct from an Intermediary Provider:

1. The ability to provide a single consistent interface (for business or technical purposes) to multiple differing Providers, and
2. The *transparent visibility* that the Broker allows into who is providing the services in the background – as opposed to Intermediary Providers that do not offer such transparency.

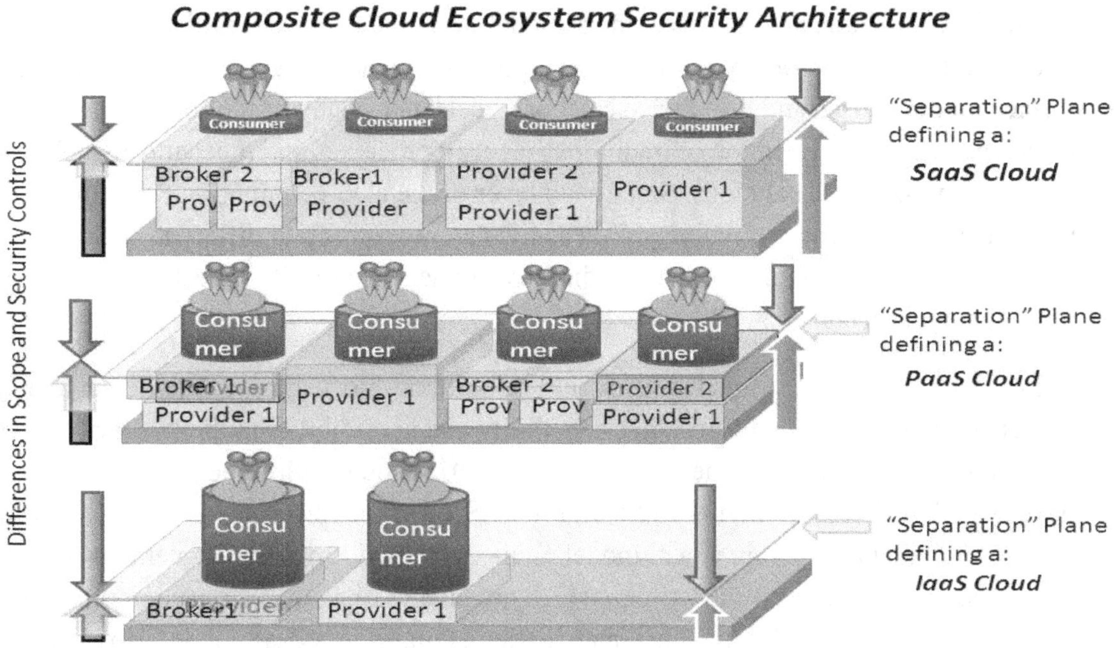

Figure 3: Composite Cloud Ecosystem Security Architecture (Courtesy of NIST)

Since the Technical Broker allows for this transparent visibility, the Consumer is aware of which key management functions are implemented by each Actor. This case is different from the case in which an Intermediary Provider is involved, since the Intermediary Provider is opaque, and the Consumer is unaware of how the key management functions are divided, when applicable, between the Intermediary Provider and the Primary Provider.

The NIST RA diagram in Figure 2 also depicts the three service models discussed earlier: IaaS, PaaS and SaaS in the 'inverted L" representations, highlighting the stackable approach of building a cloud service. Additionally, the NIST RA diagram identifies, for each cloud Actor, their general activities in a cloud ecosystem. This Reference Architecture is intended to facilitate the understanding of the operational intricacies in cloud computing. It does not represent the system architecture of a specific cloud computing system; instead, it is a tool for describing, discussing, and developing a system-specific architecture using a common framework of reference that we plan to leverage in our later discussion of key management issues in a cloud environment.

Cloud computing provides enterprises with significant cost savings, both in terms of capital expenses (CAPEX) and operational expenses (OPEX), and allows them to leverage leading-edge technologies to meet their information processing needs. In a cloud environment, security and privacy are a cross-cutting concern for all cloud Actors, since both touch upon all layers of the cloud computing Reference Architecture and impact many parts of a cloud service. Therefore, the security management of the resources associated with cloud services is a critical aspect of cloud computing. In a cloud environment, there are security threats and security requirements that differ for different cloud deployment models, and the necessary mitigations against such threats and cloud Actor responsibilities for implementing security controls depend upon the service model chosen and the service categories selected. Many of the security threats can be mitigated with the application of traditional security processes and mechanisms, while others require cloud-specific solutions. Since each layer of the cloud computing Reference Architecture may have different security vulnerabilities and may be exposed to different threats, the architecture of a cloud-enabled service directly impacts its security posture and the system's key management aspects.

For each service model, **Figure 4** below uses a building-block approach to depict a graphical representation of the cloud Consumer's visibility and accessibility to the "Security and Integration" layer that hosts the key management in a cloud environment. As the figure shows, the cloud Consumer has high visibility into the "Security & Integration" layer and has control over the key management in a IaaS model, while the cloud Providers implement only the infrastructure-level security functions (which are always opaque to Consumers). The Consumer has limited visibility and limited key management control in a PaaS model, since the cloud Provider implements the security functions in all lower layers except the "Applications" layer. The cloud Consumer loses the visibility and the control in a SaaS model and, in general, all key management functions are opaque to the cloud Consumer, since the cloud Provider implements all security functions.

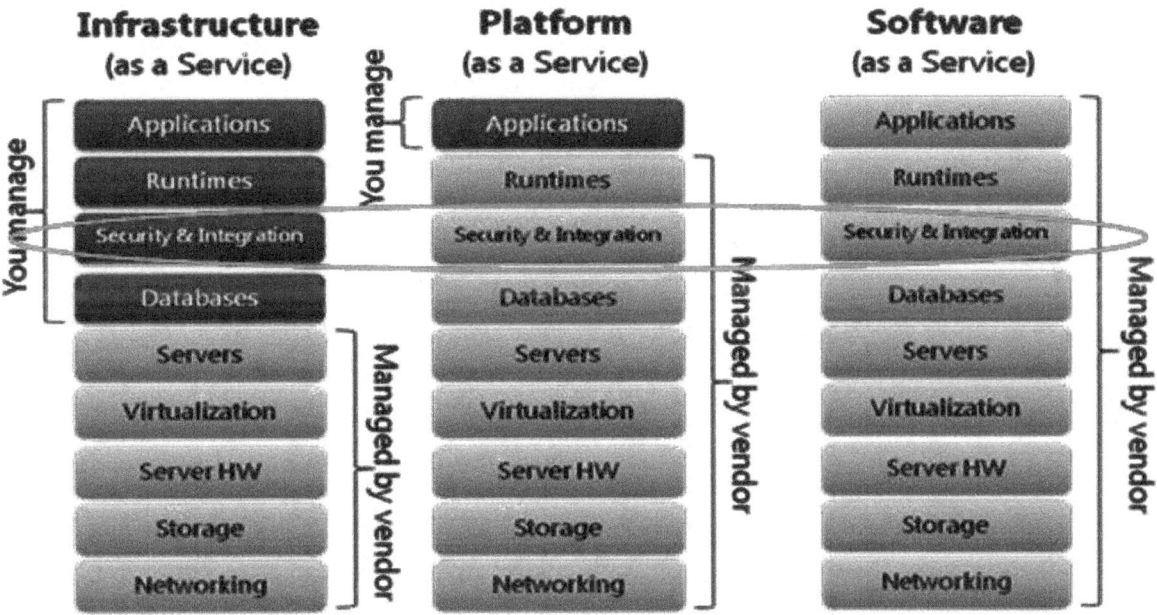

Figure 4 Cloud Service Models and Data Protection (Courtesy of CIO Research Council |CRC|)

In the following Section, we will discuss, for each service model, the Key Management challenges encountered by cloud Actors in different use cases.

3. Cryptographic Key Management Challenges in the Cloud

As stated in Section 2, the secure management of the resources associated with cloud services is a critical aspect of cloud computing. Cryptographic operations form one of the main tasks of secure management. Hence, while cloud services provide ubiquitous computing, elastic capabilities and self-configurable resources at lower costs, they also entail performing several cryptographic operations (from a cloud Consumer perspective) for the following:

- Secure Interaction of the Cloud Consumer with various services and
- Secure Storage of data generated/processed by those services.

The key management system (KMS) required to support cryptographic operations for the above functions can be complex, due to differences in ownership and control of underlying infrastructures on which the KMS and the protected resources are located. For example, though the ownership of data in cloud services rests with the cloud Consumer, the data is physically resident on storage resources controlled by the cloud Provider, and in many instances, the KMS required for managing the cryptographic keys needed to protect that data have to be run on the computing resources provided by the cloud Provider. This presents challenges to a cloud Consumer seeking to obtain the necessary security assurance from those cryptographic operations.

15

The driver for the set of cryptographic operations performed in the main cloud service models (IaaS, PaaS and SaaS) depends upon the features that constitute those services. Though there are slight variations in the feature set among different cloud Providers, it is possible to identify a core set of features. Based on these core set of features, we identify the security capabilities associated with the exercise of the features, and from the state of practices using architectural solutions for achieving those security capabilities, we derive the key management challenges for IaaS, PaaS and SaaS service types in sections 3.1, 3.2 & 3.3, respectively. *It must be noted upfront that in all architectural solutions where cryptographic keys are stored in the cloud, there is a limit to the degree of security assurance that the cloud Consumer can expect to get, due to the fact that the logical and physical organization of the storage resources are entirely under the control of the cloud Provider.*

3.1 Challenges in Cryptographic Operations & Key Management for IaaS

In the IaaS cloud type, the Consumer deploys its own computing resources in the form of virtual machines (VMs) or leases them from the cloud Provider. The leasing option involves checking out pre-built images offered by an IaaS cloud Provider. The VM images that are checked out must be authenticated to ensure that they are from authorized sources and have not been tampered with. After a VM is configured, it has to be launched in the cloud Provider's infrastructure to become a running VM instance. The operation of launching the VM and the subsequent lifecycle operations on the VM (such as Stop, Pause, Restart, Kill etc) are performed by the IaaS cloud Consumer through access to the management interface of the Hypervisor. Additionally, during operations or the use of cloud services, the IaaS cloud Consumer has to interact with running VM instances in a secure manner. These three operations – checking out a VM, performing lifecycle operations (including launching) on a VM instance and secure interaction with it - are performed by designated service-level administrators of the IaaS cloud Consumer. IaaS cloud service *security capabilities (SC)* that enable these operations are:

- **IaaS-SC1**: *The ability to authenticate pre-defined VM Image Templates made available by a cloud Provider for building functional, customized VM instances that meet a cloud Consumer's needs,*
- **IaaS-SC2**: *The ability to authenticate the API calls sent by the cloud Consumer to the VM Management interface of the cloud Provider's Hypervisor environment, and*
- **IaaS-SC3**: *The ability to secure the communication while performing administrative operations on VM instances*

For each of the three security capabilities identified above, possible *architectural solutions* (AS) are presented below that are based on known secure functions or protocols. The cryptographic key management challenges associated with these AS are also described and discussed.

IaaS-SC1: *The ability to authenticate pre-defined VM Image Templates made available by a cloud Provider for building functional, customized VM instances that meet a cloud Consumer's needs (Server Authentication Mechanism).*

Architectural Solution:

When leasing VMs from IaaS Providers, cloud Consumers are concerned that the VM image templates being checked out might not be authentic. To mitigate this concern, the templates can be digitally signed by the cloud Provider. The private key of a public/private key pair that is used to sign the VM image templates should be securely stored by the Provider and protected while in use (e.g., using FIPS 140-2 validated cryptographic module). The Provider needs to make the corresponding public key available to the Consumer in an authenticated manner (e.g., using an out-of-band means or using a public key certificate). Alternative means of assuring the integrity of the VM are: a) the use of a cryptographic hash function (secure hash function), such as SHA-256 computed over the VM code, which Consumers should re-compute and verify against the value obtained using an out-of-band means; b) the use of cryptographic message authentication code (MAC) mechanisms (i.e., HMAC or a block-cipher-based MAC) using a cryptographic algorithm and a secret shared by the Provider and the Consumers.

Key Management Challenges:

The authentication of the VM templates using one of the cryptographic techniques referred above (i.e., digital signature, cryptographic hash function, or message authentication code) entails the bootstrapping problem and hence, requires a comprehensive security analysis, rather than just an examination of the key management challenge. Appendix A provides this analysis for the three possible cryptographic techniques for achieving IaaS-SC1 and a possible solution.

IaaS-SC2: *The ability to authenticate the API calls sent by the cloud Consumer to the VM Management interface of the cloud Provider's Hypervisor environment.*

Architectural Solution:

Although the responsibility for configuring the VMs lies with a cloud Consumer, an IaaS cloud Provider can implement functionality whereby the VM Management Interface of the Hypervisor only accepts and executes authenticated API calls. Cloud Consumers need to generate or possess a public/private key pair that will be used for signing the calls submitted to the VM Management interface. The public key needs to be bound to the Consumer's identity in a public key certificate signed by a trusted authority. The certificate is then made available to the VM Management Interface of the Hypervisor to verify the signature of the calls submitted by the Consumer to the VM instance. An alternative approach is to provide the capability for the cloud Consumer to set up a secure session with the VM Management interface using either SSH (refer IaaS-SC3) or TLS (refer IaaS-SC4).

Key Management Challenge:
Cloud Consumers need to secure the private key of the public/private key pair that is used to sign the VM Management commands on their system (both at rest and while in use).

IaaS-SC3: *The ability to secure the communication while performing administrative operations on VM instances.*

Architectural Solution:
The service-level administrators of the IaaS Consumer need root/administrator access to running VM instances deployed or leased by that Consumer. A typical mechanism deployed to secure this access is Secure Shell (SSH) that provides a framework for public/private (asymmetric) keys or password-based client authentication techniques. A public/private key technique requires the cloud Consumer to generate a public/private key pair and then associate the public key with the Consumer's account in the VM instance. The task of a VM recognizing the Consumer as the owner of the companion private key is accomplished by appending the public key to the authorized keys file in the VM instance that can support SSH login through protocols such as File Transfer Protocol (ftp), Secure Copy Protocol (scp), or console commands. Thus, SSH can be used to enable the VM instance to authenticate the Consumer using cryptographic means. Further details of the SSH protocol are described in Internet RFC 4253. This strong cryptographic authentication prevents anonymous connection attempts to the VM instance, as well as preventing authentication attacks (such as password guessing). Moreover, the SSH protocol permits asymmetric keys to be used to perform an authenticated ephemeral Diffie-Hellman (DH) key establishment. The symmetric session keys calculated during this process are used to encrypt the payload and to generate hash-based message authentication codes, thus providing both confidentiality and integrity security services. When SSH is used, not only is the administrator authenticated, but all the commands, responses, and payload are protected in both directions (Consumer ←→ VM) from eavesdropping and against undetected modifications, and are cryptographically authenticated.

Key Management Challenges:
Cloud Consumers need to secure the private key of the public/private key pair that is used to authenticate themselves, using the best enterprise security mechanisms. It is important to note that, the Diffie-Hellman keys and the derived session keys are ephemeral and generated or calculated on-the-fly. Thus, these keys do not require persistent storage, and hence, their key management is not an issue.

After the service-level administrator of the cloud Consumer authenticates pre-defined VM Images provided by the cloud Provider and checks them out (using capability IaaS-SC1), customizes them to its requirements, launches them securely in the hypervisor environment (using IaaS-SC2) of the cloud Provider and performs configuration maintenance through secure interaction with the launched VM instances (using capability IaaS-SC3), the application-level administrator of the cloud Consumer installs and configures various servers (web servers, Database Management servers, etc.), application execution environments (e.g., Java VMs and Java run time modules) and application executables (and in some instances, source codes, as well) on those VM instances. Although the application-level administrators do not configure VM instances (such as allocation/resizing of virtual memory, CPU cores, or virtual disks), they need to setup secure sessions with VM instances prior to being authenticated. Hence, in most practical situations, the same service-level administrators of the cloud Consumer play the role of application-level administrators as well. The administrators use the same SSH technique and keys for secure application-level administration.

After applications are up and running on their leased VMs, the application users of an IaaS cloud Consumer would like to interact with these applications securely (through setting up secure sessions and strong authentication) and exercise the various application features – depending upon the set of assigned permissions or by assuming their assigned roles (which provide the permissions). Finally, there is the need for Data Storage services for all categories (service-level administrators, application-level administrators and application users) of IaaS Consumers. The data storage services required may span different types of data, such as: (a) Static Data – applications' source code, Reference data used by applications, Archived data and Logs, and (b) Application data – generated and used by applications. The application data in turn could be either Structured (e.g., Database data) or Unstructured (e.g., files from social feeds).

The challenges in the secure interaction of the application users (as opposed to application-level administrators) of IaaS cloud Consumers with IaaS cloud services (both main services, such as executing the applications on VM instances, as well as auxiliary services such as data storage) are:

- IaaS-SC4: *The ability to secure the communication with application instances running on VM instances for application users during cloud-service usage,*
- IaaS-SC5: *The ability to securely store static application support data securely* (data not directly processed by applications),
- IaaS-SC6: *The ability to securely store application data in a structured form* (e.g., relational form) securely using a Database Management System (DBMS), and
- IaaS-SC7 *The ability to securely store application data that is unstructured*

    IaaS-SC4: *The ability to secure the communication with application instances running on VM instances for application users during cloud service usage.*

Architectural Solution:
Application users (clients) generally interact with services by setting up a secure session (which can provide both confidentiality and integrity) with application (service) instances (e.g., Web

server or DBMS server instances). The most common technology employed is the Transport Layer Security (TLS) protocol. TLS, just like SSH described earlier, can be used to enable the service instance and client to authenticate each other using a cryptographic means (as described in Internet RFC 5246), as well as to set up secure session keys for encrypting/decrypting and for generating message authentication codes.

Key Management Challenges:
The secure session requires the presence of an asymmetric key pair (private and public keys) for a service instance and an optional key pair on the client side, as well. The client-side private key can be managed by an enterprise key management system, and the server-side private key has to be managed by a key management system run by the IaaS cloud Provider.

IaaS-SC5: *The ability to securely store static application support data securely.*

Architectural Solution:
To support applications running on leased VM instances, IaaS cloud Consumers need secure storage services to store relatively static data such as applications' source code, reference data used by applications, preferred VM Images, and archived data and Logs. These types of data are different from data generated, processed, and stored directly by the application. To store the former type of data, the cloud Providers offer a file-storage service.

Key Management Challenge:
The data that is not processed by or written to by applications can be encrypted at the cloud Consumer site before being uploaded to the cloud Providers file storage service. Hence, encryption keys (generally, symmetric keys) needed for encrypting the data at the cloud Consumer site and are under its administrative control and can thus be secured using enterprise key management solutions.

IaaS-SC6: *The ability to securely store application data in a structured form securely:* To store structured data generated by applications running on its VM instances, the IaaS cloud Consumer needs to subscribe to a Database service (generally a relational service offered by the Provider as an adjunct to its IaaS offering). The cloud Consumer subscribing to this service is generally provided with a DBMS instance with the ability to custom configure the instance to suit its business and security needs. The options available to provide confidentiality protection for data managed by the DBMS instance and the associated key management challenge are described below:

Architectural Solution-TDE: (Transparent/External Encryption):
Use the native encryption function that is provided as a feature within the DBMS engine or use a third party tool. This feature is called Transparent Data Encryption (TDE) and is a technique

similar to storage-level encryption (the encryption engine operates at the I/O level and encrypts data just prior to being written to disk). The whole database is protected with a single Database Encryption Key (DEK) that is itself protected by more complex means, including the possibility of using a Hardware Security Module (HSM). Since TDE performs all cryptographic operation at the I/O level within the database system, there is no need to modify the application logic or the database schema.

Key Management Challenge:
Since the IaaS cloud Consumer has administrative control of the subscribed DBMS instance, it has control over the DEK as well. Since encryption is taking place at the I/O level, the DEK has to reside close to the storage resources designated for storage of the database data, and hence, the cloud Consumer has no option other than storing the DEK in the same cloud where the DBMS instance is running. Although there are TDE implementations that offer column and table-level granularity for encryption, the most common usage is for storage-level encryption, and hence, the implementation cannot be configured to provide different encryption keys to different users based on their permission set (or assigned role).

Architectural Solution -ULE: (Database level encryption or user-level encryption)
Under this feature, users can choose to encrypt data at the column level, table level or even a set of data files corresponding to multiple tables or indexes.

Key Management Challenge:
This solution requires the use of a different encryption key for different database objects. An additional service is required (e.g., by a Security Server) that will map the set of session permissions of the user (based on the roles assumed) to the set of keys, and then make a call to a KMS to retrieve the required set of keys from key storage. For better security, the security server, the KMS, and (persistent) key storage should be run in a cloud that is different from the DBMS instance or should be run on-premise by the cloud Consumer. The security server and KMS perform the role-to-key mapping and key retrieval functions, respectively, based on the authenticated credentials of the DBMS user. However, during a user's session (for key usage), the keys remain in a cache of the memory space created for the user session in the same cloud as the DBMS instance. The added challenge of retrieving the key from the KMS and providing it securely to the application running in the cloud Provider space also needs to be dealt with. One can argue that once the secure session with the DBMS application in the cloud is established, this security challenge is trivial. Alternatively, the cloud Consumer can run the security server and the KMS in the same cloud as the DBMS application. This latter approach leaves the sensitive data vulnerable to access by the cloud Provider Administrators unless additional security measures are taken.

IaaS-SC7: *The ability to store unstructured application data securely:* This operation requires storage-level encryption similar to *Transparent/External encryption* (**Architectural Solution-1: (Transparent/External Encryption**), and hence, the same key management challenges apply.

3.2 Challenges in Cryptographic Operations and Key Management for PaaS

The objective of a Platform as a Service (PaaS) offering is to provide a computational platform and the necessary set of application development tools to Consumers for developing or deploying applications. Although the underlying OS platform on which the development tools are hosted is known to the Consumer, the Consumer does not have control over its configuration functions and thus the resulting operating environment. Consumers interact with these tools (and associated data, such as development libraries) to develop custom applications. Consumers may also need a storage infrastructure to store both supporting data and application data for testing the application functionality. PaaS cloud service *security capabilities (SC)* that enable these operations are:

- PaaS-SC1: *The ability to set up secure interaction with deployed applications and/or development tool instances,*
- PaaS-SC2: *The ability to securely store static data* (data not directly processed by applications),
- PaaS-SC3: *The ability to securely store application data in a structured form* (e.g., relational form*) using a Database Management System* (DBMS), and
- PaaS-SC4: *The ability to securely store application data that is unstructured.*

The operations involved in exercising the above capabilities (PaaS-SC1 through PaaS-SC4) are identical to the operations involved in exercising capabilities IaaS-SC4 through IaaS-SC7, respectively. Therefore, the same cryptographic key management challenges apply.

3.3 Challenges in Cryptographic Operations & Key Management for SaaS

SaaS offerings provide access to applications hosted by the cloud Provider. An SaaS cloud Consumer would like to interact with these application instances securely (through setting up secure sessions and strong authentication) and exercise the various application features, depending upon the set of assigned permissions or by assuming their assigned roles (which provide the permissions). In addition, some SaaS Consumers would also like to store the data generated/processed by those applications in an encrypted form for the following reasons: (a) to prevent exposure of their corporate data, due to loss of the media used by cloud Providers; (b) surreptitious viewing of their data by an SaaS co-tenant or by a cloud Provider administrator. Though the former feature (secure interaction with application) is provided by the SaaS Providers, the second feature (storing data in an encrypted form) currently has to be provided entirely by the SaaS Consumer. The typical set of security capabilities (whether provided by an SaaS service or not) are:

- SaaS-SC1: *The ability to set up secure interaction with an application, and*
- SaaS-SC2: *The ability to store application data* (structured or unstructured) *in an encrypted form.*

The operations involved in exercising the SaaS-SC1 capability are identical to the operations involved in exercising the IaaS-SC4 capability. Therefore, the same cryptographic key management challenges apply.

SaaS-SC2: *The ability to store application data* (structured or unstructured) *in an encrypted form.*

There are two operational scenarios here. If all fields in the database need to be encrypted, then the encryption capabilities have to reside with the cloud Provider because of the sheer scale of operation (see Architectural Solution – DVE below for description). On the other hand, if each cloud Consumer wants selective encryption of a subset of fields, and since this subset varies with each Customer, all encryption operations have to take place at the client (cloud Consumer) end (see Architectural Solution – GTE). The key management challenges for each of the two options are discussed below after a brief description of the associated architectural solution.

Architectural Solution-DVE (Encryption of Entire Database):
For efficient encryption and storage of application data, SaaS cloud Providers divide the physical storage resources into logical storage chunks called disk volumes and assign different encryption keys over sets of disk volumes (e.g., assign an encryption key for two or three disk volumes).

Key Management Challenge:
Since all the encryption keys are under the control of the SaaS cloud Provider, this architectural solution does not provide assurance to the Consumer against the insider[3] threat unless additional measures are taken. Also, it is possible that data belonging to different Consumers reside on a single disk volume and is protected by a common encryption key, providing no cryptographic separation of the data belonging to different cloud Consumers. Furthermore, the sheer volume of data stored in large SaaS cloud offerings requires a large number of keys, thus necessitating the need for the management of hundreds of symmetric encryption keys, possibly using multiple key management servers. If the key management function is carried out using an HSM, then it may require the creation and maintenance of multiple HSM partitions.

---

[3] That is, cloud Provider Administrator

Architectural Solution-GTE (Selective Encryption of Database fields):

For selective encryption of certain set of fields chosen by the Consumer (the selection of the set based on each Consumer's business requirements), an encryption gateway (generally running as an appliance) is usually employed inside the cloud Consumer's enterprise network. Architecturally, the gateway is located between the SaaS client application and SaaS cloud application (hosted by cloud SaaS Provider) and acts as a reverse proxy server that monitors all incoming and outgoing application traffic (e.g., HTTP, SMTP, SOAP and REST). The outgoing payload in this context will usually be the data that needs to be sent to the SaaS cloud application for storage. The gateway being configured with rules for encrypting different data items, encrypts or tokenizes the data in real time and forwards the modified data to the SaaS cloud application. Similarly, encrypted or tokenized data retrieved and returned by the SaaS cloud application is converted again, in real time, into clear text prior to being displayed by the SaaS client application. This encryption scheme thus requires no change either to the SaaS cloud Provider application or to the SaaS cloud Consumer's client application. Furthermore, all application functionality can be exercised normally since the encryption/decryption process performed by the encryption gateway is Format and Function-Preserving. Thus, the encryption gateway is the solution adopted under the following scenario:

- The SaaS cloud Consumer needs selective encryption of certain fields and hence all the processing (from the application functionality point of view) as well as encryption of those fields occurs at the Consumer side and the DBMS instance at the cloud is used just for storage (as opposed to computational processing) as far as those fields are concerned.
- The values in fields marked for encryption thus are in encrypted form at all times in the cloud (both during application processing in the cloud and storage in the cloud)
- Data in clear text is visible only to authorized clients using SaaS client application to interact with the SaaS cloud application through the encryption gateway

Key Management Challenge:

The encryption gateway may use a single key or different cryptographic keys for encrypting/decrypting different selected fields of the application. Irrespective of the number of cryptographic keys used, since the encryption gateway resides within the enterprise network perimeter, all cryptographic keys are fully under the control of the SaaS cloud Consumer and, as such, protected using in-house enterprise key management policies and practices.

Appendix A – Security Analysis of Cryptographic Techniques for Authenticating VM Templates in the Cloud

When leasing VMs from cloud Providers, cloud Consumers are concerned that the VM templates being checked out might not be authentic. To mitigate this concern, the following are some possible techniques:

1. A Digital Signature on the VM template,
2. The use of a Cryptographic Hash function,
3. The use of a Keyed Message Authentication Code, or
4. The use of cloud Provider Environment Discretionary Access Control.

Each of these techniques is described and analyzed below. Note that there are numerous variations for each technique and several other techniques, but these techniques were chosen to illustrate how to go about performing security analysis. Also note that, based on the cloud computing paradigm, it is assumed that the cloud Consumer will not download the VM template for authentication in the Consumer's Enterprise environment. Rather, the authentication will be performed in the Provider environment in which the VM is going to execute.

A.1. VM Template Authentication using Digital Signature

As Figure A-1, illustrates, the cloud Provider signs the VM template using the cloud Provider's private key once the VM template has been created. The signing function needs to be performed only once when the VM template is created.

Every time that a cloud Consumer checks out a VM template, he/she can verify the digital signature on the VM template using the public key of the cloud Provider. The cloud Consumer supplies the public key to the verification engine as illustrated in Figure A-1.

This approach has the advantage that the cloud Provider is able to create and modify multiple VM templates, and all cloud Consumers can verify the source and integrity of the VM template via a digital signature verification. It also has the advantage of simplified key management. All that is required are the following: a) the cloud Provider needs to create a single public/private signature key pair and protect the private key from unauthorized use and from unauthorized disclosure, b) the cloud Provider needs to provide the public key in a trusted manner[4] to each cloud Consumer; and c) the cloud Consumer needs to protect the public key from undetected, unauthorized modification.

---

[4] This can be easily accommodated using physical means during contract signing.

The approach has some disadvantages as well. While on the surface, the approach seems highly secure, there are several security concerns with it:

1. First of all, how does the cloud Consumer communicate securely with the verification engine to provide the public key and to obtain the verification results. Let us assume that the cloud Consumer can establish a secure session using TLS or SSH.
2. Then the question becomes: how does the cloud Consumer trust the verification engine running in the cloud Provider. If the cloud Consumer cannot trust or authenticate the verification engine, it has no basis to trust the response from the verification engine regarding the VM template signature verification.
3. Furthermore, whatever means the cloud Consumer uses to establish trust in the verification engine, why not use the same means to trust the VM template and forego the extra step of having to first establish trust in the verification engine?

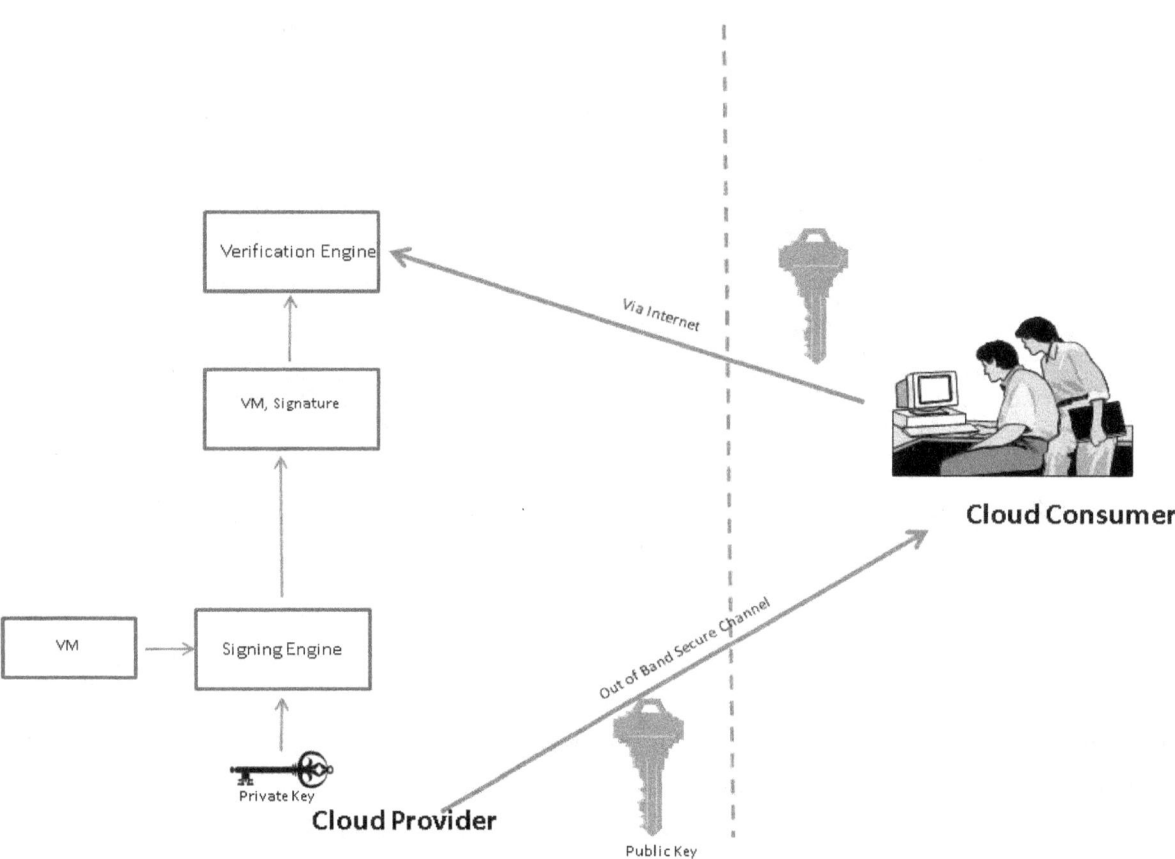

Figure A.1: VM template Authentication Using Digital Signatures

A.2. VM Template Authentication using Cryptographic Hash Function

Another technique of assuring the integrity of the VM template is by using a cryptographic hash function, such as SHA-256, to compute a hash value on the VM template, and the Consumers obtaining the hash value using an out-of-band means as illustrated in Figure A-2.

The approach has the advantage of requiring no key management. However, the hash value of the VM template needs to be provided to the consumers using means that assure its integrity and source (e.g., physically). The cloud Consumer provides this hash value for comparison during VM template authentication.

The approach has several disadvantages. Some of the disadvantages are common to those for digital signatures:

1. This approach has the limitation that each time the VM template `is modified, a new hash value needs to be promulgated using a secure, out-of-band means.

2. The approach has the limitation that each VM template hash value needs to be promulgated using secure, out-of-band means. One can assume that the cloud will have multiple VM templates.

3. Just like the digital signature, this approach does not solve the problem of the cloud Consumer communicating securely with the verification engine to provide the hash value and obtaining the verification results. Let us assume that the cloud Consumer can establish a secure session using TLS or SSH.

4. Then the question becomes: how does the cloud Consumer trust the verification engine running in the cloud Provider. If the cloud Consumer cannot trust or authenticate the verification engine, it has no basis to trust the response from the verification engine regarding the VM template verification.

5. Furthermore, whatever means the cloud Consumer uses to establish trust in the verification engine, why not use the same means to trust the VM template and forego the extra step of having to first establish trust in the verification engine?

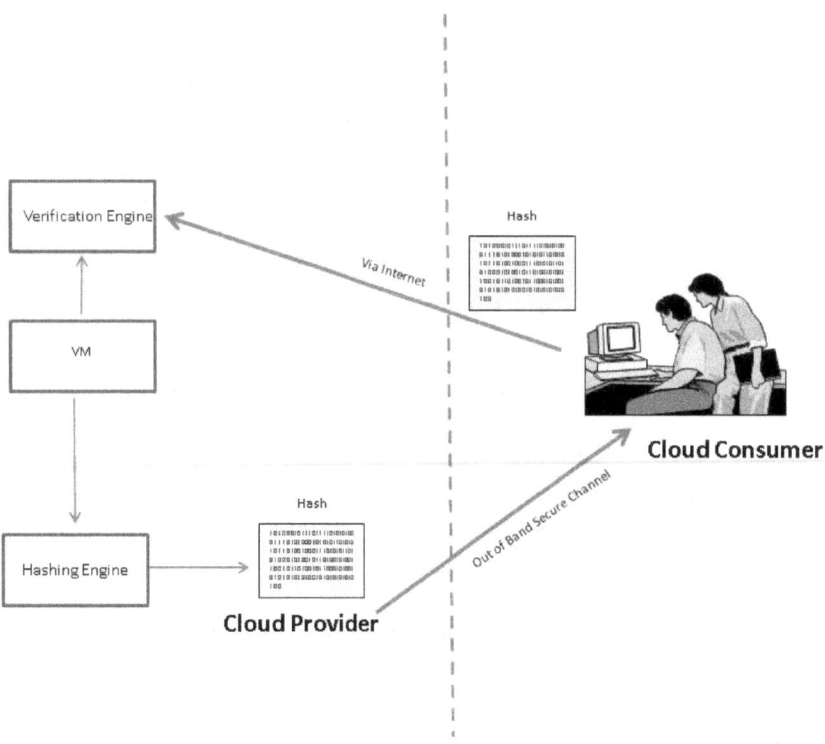

Figure A.2: VM template Authentication Using Cryptographic Hash

A.3. VM Template Authentication using Message Authentication Code (MAC)

As illustrated in A-3, another approach is to use a MAC. A MAC is calculated using a cryptographic function, such as a keyed hash function or a mode of operation for a symmetric block cipher algorithm, that produces a message authentication code using a secret shared by the Provider and the Consumers.

The approach has the advantage of the cloud Provider being able to create and modify multiple VM templates and all cloud Consumers being able to verify the source and integrity of the VM template via MAC verification. It also has the advantage of simplified key management. All that is required are the following: a) the cloud Provider needs to create a single secret key and protect it from unauthorized use and from unauthorized disclosure; b) the cloud Provider needs to provide to each cloud Consumer with the secret key in a secure manner[5]; and c) the cloud Consumer needs to protect the secret key from unauthorized disclosure.

The approach has several disadvantages. Some of the disadvantages are common to those for using digital signatures:

---

[5] This can be easily accommodated using physical means during contract signing.

1. Unless the secret key is unique per Consumer, this approach is vulnerable to one Consumer modifying a VM template to compromise another Consumer. Having unique keys for each Consumer will increase a cloud Provider's key management challenge

2. Just like the use of a digital signature, this approach does not solve the problem of the cloud Consumer communicating securely with the verification engine to provide the secret key and to obtain the verification results. Let us assume that the cloud Consumer can establish a secure session using TLS or SSH.

3. Then the question becomes: how does the cloud Consumer trust the verification engine running in the cloud Provider. If the cloud Consumer cannot trust or authenticate the verification engine, it has no basis to trust the response from the verification engine regarding the VM template authentication.

4. Furthermore, whatever means the cloud Consumer uses to establish trust in the verification engine, why not use the same means to trust the VM template and forego the extra step of having to first establish trust in the verification engine?

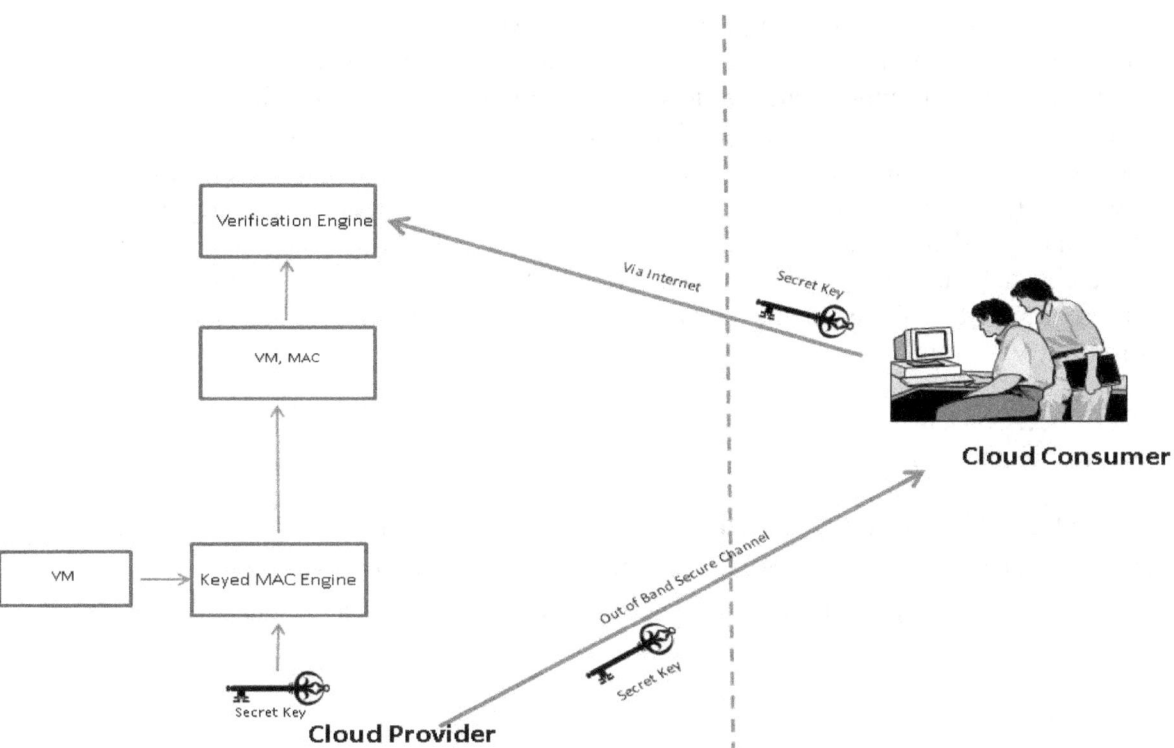

Figure A.3: VM template Authentication Using MAC

A.4. VM Template Authentication Based on Cloud Provider Discretionary Access Control

Under this approach Consumers obtain the VM template from a location that can be modified by the Provider only (i.e., the VM template is protected using discretionary access controls). Though this form of authentication is not a cryptographic technique, we have included this for completeness as a possible approach for VM template authentication.

A.5. Conclusion

In conclusion, one can see from our higher-level security analysis of the possible cryptographic techniques for authenticating VM templates, that none of them solve the twin problem of establishing trust in the VM template, as well as in the verification engine. Hence, our suggested solution for VM template authentication is:

1. The cloud Consumer should use SSL or SSH to establish a secure session with the VM template integrity verification engine.

2. The application instance housing the VM integrity verification engine needs to be configured to run as a secure appliance on a specially hardened VM. The verification engine should also include appropriate public keys, secret keys, and/or hash values, depending on the VM template authentication technique chosen by the cloud Provider. Note that this approach obviates the need for a secure, out-of-band channel between the cloud Provider and the cloud Consumer. This approach also allows the cloud Provider to change keys, algorithms, authentication method and/or a VM template without having a secure, out-of-band channel with the cloud Consumer. Note that a cloud Provider may use different cryptographic techniques (digital signatures, cryptographic hash, or MAC) to protect different VM templates.

3. The cloud Consumer should check out any VM template, and authenticate the VM template and launch the VM.

The advantage of having a verification engine as opposed to having a VM template under discretionary access control is the added flexibility for the cloud Provider to only secure the verification engine using discretionary access control, as opposed to a myriad of VM templates.

Appendix B - Bibliography

[1]   F. Liu, J. Tong, J. Mao, R. Bohn, J. Messina, L. Badger, and D. Leaf, NIST Cloud Computing Reference Architecture (NIST SP 500-292), National Institute of Standards and Technology, U.S. Department of Commerce (2011). http://www.nist.gov/customcf/get_pdf.cfm?pub_id=909505

[2]   P. Mell and T. Grance, The NIST definition of cloud computing (NIST SP 800-145), National Institute of Standards and Technology, U.S. Department of Commerce
(2011) http://csrc.nist.gov/publications/nistpubs/800-145/SP800-145.pdf

[3]   L. Badger, D. Berstein, R. Bohn, F. de Valux, M. Hogan, J. Mao, J. Messina, K. Mills, A. Sokol, J. Tong, F. Whiteside, and D. Leaf, US government cloud computing technology roadmap volume 1: High-priority requirements to further USG agency cloud computing adoption (NIST SP 500-293, Vol. 1), National Institute of Standards and Technology, U.S. Department of Commerce (2011).
http://www.nist.gov/itl/cloud/upload/SP_500_293_volumeI-2.pdf

[4]   L. Badger, R. Bohn, S. Chu, M. Hogan, F. Liu, V. Kaufmann, J. Mao, J. Messina, K. Mills, A. Sokol, J. Tong, F. Whiteside, and D. Leaf, US government cloud computing technology roadmap volume II: Useful information for cloud adopters (NIST SP 500-293, Vol. 2), National Institute of Standards and Technology, U.S. Department of Commerce (2011). http://www.nist.gov/itl/cloud/upload/SP_500_293_volumeII.pdf.

[5]   L. Badger, T. Grance, R. Patt-Corner, and J. Voas, Cloud Computing Synopsis and Recommendations (NIST SP 800-146), National Institute of Standards and Technology, U.S. Department of Commerce (2012). http://csrc.nist.gov/publications/nistpubs/800-146/sp800-146.pdf

[6]   W. Jansen and T. Grance, Guidelines on Security and Privacy in Public Cloud Computing (NIST SP 800-144). National Institute of Standards and Technology, U.S. Department of Commerce (2011). http://csrc.nist.gov/publications/nistpubs/800-144/SP800-144.pdf.

[7] Secure Shell (SSH) Transport Layer Protocol, http://www.ietf.org/rfc/rfc4253.txt

[8] The Transport Layer Security (TLS) Protocol Version 1.2, http://tools.ietf.org/html/rfc5246

[9] Internet Security Glossary, Version 2,   http://tools.ietf.org/rfc/rfc4949.txt

[10] F.Bracci, A.Corradi and L.Foschini, Database Security Management for Healthcare SaaS in the Amazon AWS Cloud, IEEE Computer, 2012.

[11] Understanding and Selecting a Database Encryption or Tokenization Solution, http://securosis.com

[12] Best Practices in Securing Your Customer Data in Salesforce, Force.com, and Chatter, http://www.ciphercloud.com

www.ingramcontent.com/pod-product-compliance
Lightning Source LLC
Chambersburg PA
CBHW082032190526

45166CB00017B/3296